Insects and *Spiders*

Bugs

Shane F McEvey
for the Australian Museum

This edition first published in 2002 in the United States of America by Chelsea House Publishers, a subsidiary of Haights Cross Communications.

Chelsea House Publishers
1974 Sproul Road, Suite 400
Broomall, PA 19008-0914

The Chelsea House world wide web address is www.chelseahouse.com

Library of Congress Cataloging-in-Publication Data Applied for.

ISBN 0-7910-6595-2

First published in 2001 by
Macmillan Education Australia Pty Ltd
627 Chapel Street, South Yarra, Australia, 3141

Copyright © Australian Museum Trust 2001
Copyright in photographs © individual photographers as credited

Edited by Anna Fern
Text design by Nina Sanadze
Cover design by Nina Sanadze
Australian Museum Publishing Unit: Jennifer Saunders and Catherine Lowe
Australian Museum Series Editor: Deborah White

Printed in China

Acknowledgements
Our thanks to Martyn Robinson, Max Moulds and Margaret Humphrey for helpful discussion and comments.

The author and the publisher are grateful to the following for permission to reproduce copyright material:

Cover: A harlequin bug, photo by Densey Clyne.

Australian Museum/Nature Focus, pp. 26, 27; B. L. Grainger/Nature Focus, p. 15 (top); Bill Belson/Lochman Transparencies, pp. 6–7, 11 (bottom left); Brian Chudleigh/Nature Focus, p. 24 (top); C. Andrew Henley/Nature Focus, pp. 11 (bottom right), 18 (bottom), 19 (top, middle and bottom left); Densey Clyne/Mantis Wildlife, pp. 4, 15 (bottom), 16 (bottom), 19 (bottom right), 23 (bottom left and right); Diane Norris/Nature Focus, p. 29; Dominic Chaplin/Nature Focus, p. 14 (top); Jiri Lochman/Lochman Transparencies, pp. 10 (bottom), 11 (top), 12 (top and bottom), 13 (middle), 16 (top), 18 (top), 22 (bottom), 25 (top), 28 (bottom); Michael Cermak/Nature Focus, p. 5 (right); Michael Terry/Nature Focus, p. 17 (top); Mike Braham/Lochman Transparencies, p. 28 (top); Pavel German/Nature Focus, pp. 8 (bottom), 9, 10 (top), 13 (top), 23 (middle), 25 (bottom); Peter Marsack/Lochman Transparencies, pp. 8 (top), 13 (bottom), 14 (bottom), 17 (bottom), 20, 21 (all), 22 (top); Steve Wilson/Nature Focus, pp. 5 (left), 24 (bottom), 30; T. & P. Gardener/Nature Focus, p.15 (middle); Wade Hughes/Lochman Transparencies, p. 23 (top).

Contents

What are bugs? 4

Bug bodies 6

Where do bugs live and what do they eat? 10

Bugs that live in deserts and dry habitats 12

Bugs that live in forests and wet habitats 14

How bugs communicate and explore their world 16

The life cycle of bugs 18

Predators and defenses 22

Weird and wonderful bugs 24

Collecting and identifying bugs 26

Ways to see bugs 28

Bugs quiz 30

Glossary 31

Index 32

Glossary words

When a word is printed in **bold** you can look up its meaning in the Glossary on page 31.

What are bugs?

Bugs are insects. Insects belong to a large group of animals called invertebrates. An invertebrate is an animal with no backbone. Instead of having bones, bugs have a hard skin around the outside of their bodies that protects their soft insides.

Bugs have:
- six legs
- four wings
- two **antennae**
- two eyes
- a sucking beak
- many breathing holes on the sides of their bodies.

Did you know?

Many people call all kinds of insects 'bugs' but this is wrong. 'Bug' is the special name given to a special kind of insect.

The insect pictured here is a typical bug.

What makes bugs different from other insects?

Bugs have piercing and sucking **mouthparts**.

Scientists have given a special name to all bugs. They are called **Hemiptera**.

Kinds of bugs

All bugs can be divided into two groups:
- Bugs that have wings that are clear or transparent all over. An example of this sort of bug is a cicada.
- Bugs that have wings that are half clear and half leathery. You cannot see through the leathery part. An example of this sort of bug is a stink bug.

Cicadas are bugs with transparent wings.

Fascinating Fact

There are more than 20,000 kinds of bugs worldwide.

This bug has its wings folded over its back but you can still see that each wing is half clear and half leathery.

Bug bodies

The body of a bug is divided into three segments. These segments are called the head, the **thorax** and the **abdomen**.

Thorax

On the thorax are:
- six legs
- four wings.

thorax

abdomen

Abdomen

The abdomen is where:
- food is digested
- females produce eggs
- males produce **sperm**.

Head

On the head are the:
- antennae
- eyes
- mouth.

Did you know?

Bugs can grow very large. Some bugs are as long as ten centimeters (4 inches).

head

This is a shield bug **nymph**. Nymphs are young bugs that are still growing into adult bugs. Their wings are not yet formed.

7

The head

On the head of an adult bug are the mouth, eyes and antennae.

Mouth

The mouth of a bug is a hard, sharp beak. The bug uses its beak to pierce its food like a needle and then suck up the juices. When the bug is not feeding, its beak is usually tucked away under its body.

Eyes

Bugs have compound eyes. This means that each eye is made up of lots of tiny eyes packed together.

eyes

beak

This assassin bug has a drop of liquid on the end of its beak.

Antennae

Bugs use their antennae to feel and smell their environment. Their antennae are usually very simple. Water bugs have short antennae that are hard to see because they are tucked away. Land bugs have longer antennae that are easy to see.

This jewel bug lives on land and has long, simple antennae that stick out from the front of its head.

The thorax

On the thorax of an adult bug are the legs, wings and breathing holes.

Legs

A bug's legs can be used for climbing, swimming, jumping, landing, cleaning itself and holding its prey. Most bugs have claws on the end of their legs.

Wings

Most adult bugs have four wings. Each wing contains a network of hard veins. These veins support the wings so that they can be used for flying. Bugs that have clear wings have their own special patterns of veins. These patterns are always the same in the same kind of bug. Scientists use these patterns to identify bugs.

Bugs that have wings that are half leathery and half clear are different to bugs with clear wings. You cannot see the veins in the leathery part and the veins in the clear part are not well formed and cannot be used to identify a bug.

Did you know?

Young bugs, called nymphs, do not have wings.

Breathing holes

Bugs breathe through tiny holes along the sides of their bodies called **spiracles**. Bugs do not breathe through their mouths.

wings

legs

This greengrocer cicada has clear wings and is using its legs to hold onto a branch. Both its legs and wings are attached to the segment of its body called the thorax.

Where do bugs live and what do they eat?

Bugs can live just about anywhere — from the tropics to very cold places, from rainforests to deserts, from mountains to the coast. A few bugs even live in rock pools on the seashore or on coral reefs. Bugs live in certain places because that is where they find their food.

Water bugs

Some bugs live on and in water. These are called water bugs. Some of the places water bugs live include:
- the surface of ponds, lakes and streams
- wet moss
- the seashore
- damp meadows.

Some bugs that live in water breathe by using a long breathing tube at the end of their abdomen.

This is a water bug. You can see its long breathing tube extending out the end of its body. These bugs are able to fly from one pool of water to another. This bug is holding its two front legs straight out in front of it and is standing on its other four legs.

Pond skaters are bugs that live on the surface of water like on a pond's surface. These pond skaters are all feeding on a dead sugar ant that is floating on the surface.

What bugs eat

Bugs eat liquid food. Many bugs suck the blood or body juices out of other living animals. Other bugs suck the fluid out of plants. Some bugs suck both animal and plant fluids. What a bug eats depends on what kind of bug it is.

Water bugs suck the body juices from tadpoles, little fish and other insects that live in the water. Occasionally they will suck the sap from water plants.

Water striders are bugs with long legs that can walk on the surface of the water. These water striders are feeding on a drowned praying mantis.

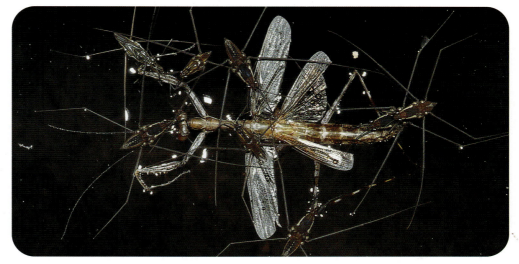

Land bugs suck sap from the roots, stems, leaves, flowers and fruits of plants. They can also suck the blood and body juices from other insects, birds and **mammals**. Land bugs smell their food. That is how they find it.

Young bugs, called nymphs, usually eat the same food as adult bugs.

This bug is feeding on the nectar of an orchid flower.

Assassin bugs are predatory bugs. This assassin bug has caught a fly and is holding it with its legs while it sucks out the fly's body juices.

Bugs that live in deserts and dry habitats

Some bugs can live in very hot and dry places like deserts. Here are some of the bugs that can live in hot, dry places.

This assassin bug has a body that looks like it is covered in sand grains. This makes the bug hard to see while it moves around on the sandy soil looking for its prey.

This shield bug nymph has striking marks on the side of its abdomen. This may help the bug to hide among the fine leaves of its food plant.

This is a cherrynose cicada. It can be found in dry eucalypt forests, especially along rivers. It sings from the highest branches during the day or at sunset.

This bug lives under eucalypt bark, where it sucks sap. Living under bark, leaves or stones is a way that insects can prevent losing water or being eaten by predators.

Did you know?

Bugs are animals.

The nymphs of some bugs make special houses to live in called **lerps**. These lerps are shaped like a dome and a nymph is living underneath each dome. Lerps are made from a waxy substance. This helps the nymph avoid drying out when the climate is hot and dry.

Bugs that live in forests and wet habitats

Lots of bugs like to live in forests. Some bugs live under water. Here are some of the bugs that live in forests and wet places.

This is the nymph of a rainforest bug. The eye spots on its back are **glands** that make special smells. This nymph feeds by sucking sap from seeds.

These redeye cicadas start singing in early November and reach their peak in late November and early December. These cicadas like smooth-barked eucalyptus trees, where they can sit and suck sap for hours at a time. Sometimes, if they are suddenly disturbed, they can fly off leaving their broken, sucking beak still stuck in the bark. They cannot grow a new beak.

This shield bug has an orange color that perfectly matches the plant it sits on. These bugs taste very bad so predators avoid them.

One of the ways that water bugs, like this water strider, manage to stay afloat is by having fine hairs on their legs that repel water. Water striders are predators and they feed on other insects.

Did you know?

Male and female harlequin bugs look very different from one another. The females are bright orange with black markings. The males are red and green. The females are also bigger than the males.

Female harlequin bugs are large and orange. This one is a female. The males are smaller and blue. Harlequin bugs were once pests of cotton plants.

How bugs communicate and explore their world

Bugs can get information about their environment in a number of ways. They can smell, feel, hear and see.

A jewel bug explores its environment.

Some bugs are active during the day and some bugs are only active at night.

Bugs have compound eyes that can see shape, movement, distance, color, and light and dark.

All bugs have antennae that they use to pick up smells from food or other bugs. They do this by sensing the special chemicals that food or other bugs release. Land bugs can also use their antennae to feel their way around the environment.

Three bronze orange bug nymphs have met on a plant. These bugs have strong smells and their antennae are touching as they cluster together.

How bugs communicate

Bugs can communicate by:
- making sounds
- releasing smells.

Lots of different kinds of bugs make sounds. Different bugs use different parts of their bodies to do this. Some rub their beak against their stomach to make a sound, some rub their leg against their side and some rub their wings against their back. Bugs use these sounds to attract other bugs.

These orange bugs are clustered on an orange tree.

Singing cicadas

Cicadas are the loudest insects. They have a very special way of making and hearing sounds. Only male cicadas sing. Inside the male cicada's abdomen is a thin piece of skin that is tight like a drum. This skin has a muscle attached to it that can pull the skin in and out. This makes a clicking sound and, because the cicada's abdomen is mostly hollow, the sound is made louder. Both male and female cicadas have a big flat plate on the underside of their bodies. This is a thin piece of skin that hears the sounds made by male cicadas.

Fascinating Fact

When the male cicada is calling, he has to loosen the skin that he uses for hearing. He does this so he cannot hear his own call, which would deafen him.

Male cicadas sing to attract female cicadas. When lots of cicadas call the sound can be deafening.

The life cycle of bugs

The whole life cycle of a bug, from egg to adult, can take a few days, many months or even many years.

Most bugs reproduce **sexually**. This means that a male and a female are needed to make new bugs. The male bug provides sperm and the female bug provides eggs. The eggs and sperm need to join together for a new bug to start growing. Males and females find each other by first being attracted to the same places. Sometimes the female attracts the male by making special smells. Sometimes the female finds the male by hearing the sounds he makes.

A group of male and female shield bugs mating.

Fascinating Fact

At certain times, some female aphids can give birth to live young without first mating with a male bug. The female is able to produce new bugs all by herself.

When the nymph molts for the last time, it turns into an adult bug. When the bug becomes an adult, it does not grow any more. How long an adult bug will live depends on what kind of bug it is. Smaller bugs can live for days or weeks. Larger bugs live longer.

An adult shield bug feeds on an insect larva.

The female lays one or many eggs. How many eggs she lays depends on what kind of bug she is. Eggs come in different shapes, depending on what kind of bug has laid them.

A female shield bug laying eggs on the leaf of a plant.

The eggs hatch into little bugs that look like adult bugs. These little bugs are called nymphs and they are often different colors to the adult they will become. The eggs are laid near or in food so that when they hatch, the nymphs will have food to eat. Some bugs give birth to live nymphs instead of laying eggs.

These eggs are hatching and tiny bright red nymphs are appearing. These nymphs are still wearing their first skin.

Fascinating Fact

The female of one kind of water bug lays her eggs on the back of the male bug. He carries the eggs stuck on his back until they hatch.

The little nymphs spend their lives eating and growing bigger. Nymphs do not have wings and cannot fly. As they grow, their skin becomes very tight until it splits. This allows the nymph to grow even bigger in a new, larger skin. This is called **molting** and it happens several times during the life of a nymph.

These growing nymphs are feeding on an insect **larva**.

19

Nymphs

Although nymphs are little versions of adult bugs, they do not have wings and are often different colors. Nymphs often like to live together.

Lerps

Some nymphs build their own houses, called lerps. Lerps protect the nymph from drying out or from being eaten by predators.

Lerps are made from a sweet, waxy substance produced by the nymph. When the nymph grows too big for its lerp it can make the lerp bigger. Different bugs make different shaped lerps. One kind of bug might make a lerp like a bubble, another bug might make a shell-shaped lerp. The nymph lives under the lerp until it becomes an adult.

Lerps can often be found on gum tree leaves. A nymph lives under each lerp, feeding on plant sap and growing bigger.

Did you know?

Some female bugs care for their eggs and nymphs by protecting them from predators.

Cicada nymphs

The cicada nymph spends its life underground. The female cicada lays her eggs on a tree. When they hatch, the nymphs crawl down the tree and burrow into the ground. The nymph stays in the ground, sucking sap from the roots of the tree and growing bigger. Some kinds of cicada nymphs stay in the ground for only one year. Others can stay for seven or more years.

When the nymph is ready to turn into an adult, it tunnels upwards. It stops just below the surface and waits until the weather is just right. Then it breaks through the surface and climbs upwards, usually on a plant.

These cicada nymphs have dug their way out of the ground and are starting to climb upwards.

The nymph then molts for the last time and a winged adult comes out. The adult stays very still while it pumps fluid into its wings to stretch them out. When their wings are hard, the cicada can fly away.

Cicada nymphs have sharp claws and this one has attached itself to the trunk of a eucalyptus tree. The nymph's skin has split down the back and the new, winged adult has begun to come out.

Fascinating Fact

One type of cicada spends 17 years underground as a nymph and only a few weeks above ground as an adult.

The new adult cicada has completely come out and is sitting on its old skin. Its wings are still soft and crumpled.

The adult has now pumped the veins in its wings full of fluid. It is sitting on its old skin while its wings harden.

21

Predators and defenses

Bugs are mostly eaten by other insects. Sometimes they are also eaten by spiders, birds, frogs, lizards and mammals.

This cicada nymph has crawled out of the ground only to end up being eaten by ants.

Bugs will attack and sometimes eat other bugs of their own kind. These bug nymphs are eating one of their brother or sister nymphs.

Did you know?

Stink bugs spray a fluid out of their bodies. This fluid stinks and, if it gets in your eye, it stings.

Protection against predators

Bugs protect themselves in a number of ways, including:
- producing strong, stinking smells
- being brightly colored, which deters predators
- flying and moving away
- living in large numbers, which reduces an individual's chance of being eaten
- being hard to see
- living inside a lerp
- looking like part of a plant.

One of the ways bugs avoid being eaten is by pretending to be something else. This leafhopper bug looks like a leaf. This makes it very hard for predators to see.

Bugs often live in large groups. One of the reasons they do this is because an individual bug in the group is less likely to be the one eaten by a predator.

Some female bugs protect their eggs until they hatch. This harlequin bug will stand guard over her eggs and protect them from being eaten.

Bronze orange bugs live on citrus trees. These bugs are able to make a strong smell to drive away any predators that try to eat them.

23

Weird and wonderful bugs

Welcome to the wonderful world of bizarre and extraordinary bugs!

Bugs in disguise

There is a bug that is shaped exactly like a rose thorn. These bugs live on roses and hide by looking like rose thorns.

Spittle bugs

Nymphs of the spittle bug make a big mass of spittle-like froth that they live in. This helps to protect them from being eaten by predators.

Spittle bug

Toe-biter bugs

Australia's largest bug lives in the water, it is sometimes called a toe-biter bug. These bugs are aggressive and will attack almost anything that moves in the water. They might even try to grab your toes if you put them into the water.

Toe-biter bug

Cochineal bugs

Red food dye called cochineal is made by crushing up a special kind of bug called the cochineal bug.

Assassin bugs

Assassin bugs inject their prey with a saliva. This kills the prey and quickly turns its insides into liquid. The bug then sucks the liquid out.

This assassin bug is feeding on a native bee.

This assassin bug gets its food by going into a spider's web and stealing the spider's prey. These bugs have a way of moving through the web without getting caught.

Fascinating Fact

Natural shellac, which is used as a furniture polish, is made by crushing up coccoid bugs.

Collecting and identifying bugs

There are so many different bugs that scientists are still discovering many new kinds. Some scientists spend years studying the thousands of bugs they have collected. If a scientist collects a bug that is unknown, they name it and describe it so other scientists can study it too.

An example of each kind of bug that has been found is kept in collections at museums. These collections are used by scientists who want to study and learn about bugs.

When scientists collect bugs, they use special equipment. They can catch them with nets or by picking them up, or they can collect them around a light at night. They need containers to put them in.

Bugs are pinned through their bodies or glued to a piece of card and placed carefully in collections. Each bug collected will have a small label attached to it that has information about where the bug was collected, when it was collected and who collected it.

Pinned bugs last longer because they do not bump around in containers which could break bits off their bodies. The pin also gives the scientist something to hold onto when they want to examine the bug.

How are bugs identified?

Bugs are identified by looking very carefully at their shape, size and color. If a bug's shape, size and color are different to all other bugs that scientists already know, then the bug is considered a new kind of bug and is given a scientific name.

What do scientists study about bugs?

After a bug has been given a name, scientists study:
- where it lives
- what it eats
- its life cycle
- what makes it grow and molt
- what makes it turn into a **pupa** or an adult
- what its natural predators are
- what poisons or pollutants kill it or interfere with its normal life cycle
- what happens if its habitat is destroyed.

Scientists are often interested in studying the changing color patterns in bugs. They want to know whether bugs are born with their color and whether the color changes when the weather is hot or cold, dry or wet.

A scientist studying bugs.

Ways to see bugs

Look on the stems and leaves of lemon trees or other **citrus** trees for different kinds of bugs. Bronze orange bugs are quite common so you might find these.

Aphids are bugs. They like to feed on many of the plants we grow in our gardens. You often see them on roses, hibiscus flowers or vegetable plants.

Look closely at ponds to see which insects skate across the surface of the water or swim just beneath it. These insects are often bugs.

Aphids

Water striders are bugs that can walk on the surface of the water.

Observing cicadas

Just about everyone has seen a cicada or heard them singing. Do the following activities to learn all about these big bugs.

- Record the sounds made by cicadas. Different types of cicadas have different songs. See if you can hear how the calls are different.

- Write down the temperature and what time of year it is when the male cicadas start singing.

- Look for nymph skins on tree trunks when you hear a cicada. Collect nymph skins and see if their claws hold on to your clothing.

- Look closely at a nymphal skin to see where it split open allowing the adult to come out.

- See if you can find the body of a dead cicada and find its piercing, sucking beak.

- Look at the nymph skins of cicadas and see how large the front legs are. These legs are for digging through the soil to get to the surface. Look how much smaller the front legs are in adult cicadas — adults do not dig in the ground.

- See how many different kinds of cicadas you can find. What colors are they?

A cicada sits on its nymph skin.

Bugs quiz

1   How do bugs protect their soft insides?

2   What sort of mouthparts do bugs have?

3   How many wings do adult bugs have?

4   What is the special name scientists have given all bugs?

5   Are all insects bugs?

6   How many body parts do bugs have?

7   Do water bugs have long or short antennae?

8   Do female cicadas sing?

9   Name two ways that bugs communicate.

10   What hatches out of a bug's egg?

11   How do bug nymphs grow bigger?

12   What lives under a lerp?

13   Where do cicada nymphs live?

14   Does every different kind of cicada have a different song?

15   Do bugs have mouths for eating solid or liquid food?

Check your answers on page 32.

These bugs can be found on citrus plants like lemon or orange trees.

Glossary

abdomen The rear section of the body of an animal.

antennae The two 'feelers' on an insect's head that are used to feel and smell. (Antennae = more than one antenna.)

aphids Small insects that suck sap from plants. Aphids are bugs.

citrus A kind of plant. Oranges, lemons, grapefruit and limes are citrus fruits.

gland Part of an animal's body that produces a fluid that has a special function or use.

Hemiptera The scientific name for bugs, a special kind of insect.

larva Caterpillars, grubs and maggots are kinds of larvae. In the life cycle of an insect the larval stage is after the egg stage and before the pupal stage. Larvae hatch out of eggs, grow and then turn into pupae. (Larvae = more than one larva.)

lerp A waxy covering produced by a bug. Bugs that make lerps live under them.

mammal A kind of animal that has a backbone and feeds milk to its young (like humans, dolphins, wombats and platypuses).

molting When an animal sheds its entire skin it molts. The process is called molting.

mouthparts Structures around the mouth that help an animal handle its food.

nymph In the life cycle of an insect (like bugs and dragonflies), eggs hatch into little nymphs that grow into adults.

pupa In the life cycle of an insect (like moths, beetles and flies), larvae turn into pupae. Adults later emerge from pupae. A cocoon is a shell around a pupa. (Pupae = more than one pupa.)

sexual reproduction When a male and female living thing combine to make more living things.

spiracle A tiny breathing hole in the side of a spider's or an insect's body.

sperm The male reproductive cell.

thorax The middle section of an animal's body.

Index

Age 21
antennae 4, 7
aphids 18, 19, 28
assassin bug 8, 11, 25

Bed bugs 10
body segments 6
breathing 4, 9
bronze orange bug 16, 17, 23, 28

Camouflage 12
cherrynose cicada 13
cicada 5, 21, 29
cochineal bug 25
collecting 26, 27, 29
color 12, 15
cotton 15

Defenses 22, 23
deserts 12

Eggs 18, 19
eyes 8

Flowers 11
food 11, 20, 21

Glands 14
growth 19

Habitats 10, 12–15
harlequin bug 15, 23
Hemiptera 5

Identifying 26, 27

Kinds 5, 29

Legs 9
lerps 13, 20
life cycle 18–21

Mating 18, 19
mimicry 23, 24

molting 18, 19, 21, 29
mouthparts 5, 8

Names 26, 27
nymph 6, 7, 19, 20

Pond skaters 10
predators 11, 22, 23

Redeye cicada 14
reproduction 18, 19

Senses 16
shield bugs 18, 19
singing 13, 14, 17, 29
size 7
sounds 13, 17, 29
spider webs 25
spittle bugs 24
stinking 14, 17
studying 26–29
sucking beak 4, 5, 8, 14

Toe-biter bugs 24

Veins 21

Water bugs 10, 15, 24
water striders 11, 15, 28
wings 4, 21
wings, clear 5, 9
wings, leathery 5, 9

Answers to quiz

1 by having hard bone-like skin around the outside 2 piercing and sucking 3 four 4 Hemiptera 5 no 6 three 7 short 8 no, only males 9 they make sounds and they release smells 10 a nymph 11 by molting 12 a bug nymph 13 in the ground 14 yes 15 liquid.